Confidently Calculated

A Guide to STEM Professional Development for Teens

Monique Sanders

Copyright © 2022 Monique Sanders

All rights reserved.

ISBN:9798364350435

DEDICATION

I dedicate the works in this book to the many young people that I have been privileged to serve over the past several years. I started my programs as a way to give back to the community and positively influence, not realizing the magnitude of impact that it would have on my own life and my foundational purpose.

CONTENTS

	Preface	1
1	Developing Skills	Pg 7
2	Assessing Your Strengths	Pg 23
3	Networking	Pg 34
4	Polish Your Professionalism	Pg 45
5	Prepare for Greatness	Pg 61
6	Develop a Plan	Pg 73
7	Stay Focused	Pg 85
8	The Mind, Body, Soul Approach	Pg 98
9	You Have the Power	Pg 106

FORWARD

"I envision leading a coalition of talented youth equipped with the confidence and knowledge to diagnose and develop solutions to critical world problems; using their experience, logical rationale and innovative perspectives."

-M. Sanders

I am Monique Magee Sanders, Chief Executive Officer and Founder of Get Stemulated, LLC., an organization designed to provide supplementary education and experimental workshops to provide exceptional Science, Technology, Engineering and Mathematics (STEM) focused curriculum. The program incorporates lessons to further promote career preparedness upon graduation for elementary, middle, and high school students.

Over the past six years, I have hosted workshops at the Boys & Girls Club, YWCO, various churches, schools and community centers. I also launched two elementary school National Society of Black Engineers (NSBE) Jr. Chapters; introducing students to the wonderful world of engineering and leadership. More recently, my company expanded its services and embarked upon project development creating a newly remodeled elementary school STEM Classroom; incorporating a makerspace and robotics stations for students to create, problem solve, and develop mechanical skills, many of the same concepts you will learn in this book.

In addition to my leadership as an entrepreneur, I have achieved a portfolio of accomplishments in the engineering, maintenance, and technical operations field; leading and managing domestic and international continuous improvement programs. In my career, I have had the opportunity to lead multi-disciplined teams in a variety of industries including agrochemical, pharmaceutical, governmental, medical device and consumer product companies such as Janssen Supply Chain Pharmaceutical Companies of Johnson & Johnson, Syngenta Crop Protection, Inc. and the U.S Coast Guards, Department of Homeland Security.

Diversity & Inclusion initiatives have played an integral part in my career, as a result I am often leading programs like Johnson & Johnson's **_WiSTEM_^{2}D** (Women in Science, Technology, Engineering, Math, Manufacturing and Design) United Program, to provide STEM career coaching, engagement and recruitment efforts to female collegiate students at the following Georgia colleges and universities including: University of Georgia, Clemson University, Spelman College, Georgia Tech, Agnes Scott College and Albany State University.

I urge readers to use this guide as a benchmark to create a plan for your educational and career journey. It is my desire that this book serves as a resource to help simplify your future planning goals with clarity so that your plans are Confidently Calculated!

Confidently Calculated

PREFACE

What is STEM?

S-T-E-M is an acronym for Science, Technology, Engineering, and Math. It is a curriculum based on educating students in that discipline. It was first introduced in 2001 by scientific administrators at the U.S National Science Foundation (NSF).

STEM classes develop the soft skill that many careers need; it encompasses hands-on, engaging subject matter idesigned to encourage critical thinking. Non-STEM fields recruit graduates with STEM degrees because they possess skills that students (who attend schools that understand the benefits of the stem in schools) from other areas have not developed. STEM skills naturally lead to more captivating performances, polished presentations, advanced problem solving skills, etc. Despite all of this, STEM scares many students because they perceive it to be too tricky and tragically tedious – This is not the case.

Why Is STEM Important?

The global economy is changing. Current jobs are disappearing due to automation, and new jobs are emerging daily due to technological advances.

The continual advances in technology are changing how students learn, connect and interact daily. Skills developed by students through STEM provide them with the foundation to succeed at school and beyond.

Employer demand for STEM qualifications and skills will continue to increase. Currently, 75 percent of jobs in the fastest-growing industries require workers with STEM skills. The modern workforce needs people who can adapt to a changing workplace to be competitive.

STEM empowers individuals with the skills to succeed and adapt to this changing world.

Benefits of STEM In High School

According to the U.S. Department of Education, young people need the "skills to solve problems, make sense of information, and knowledge to gather and evaluate evidence to make decisions." Indeed, as future leaders, they will face increasingly complex problems, but a working knowledge of STEM can help.

High school students can now enroll in robust STEM courses, expanding their understanding of the world as they develop important life skills.

1. IMPROVED PROBLEM-SOLVING ABILITIES

Typically, STEM courses rely on hands-on activities to teach a logical, systematic approach to problem-solving; as a result, the projects and interactive assignments ignite curiosity.

Additionally, STEM courses strengthen critical thinking as students learn to analyze a problem sequentially: identifying the root problem, proposing a hypothesis, testing theories, recording data, evaluating trends, and positing a sustainable solution.

More importantly, STEM courses can be fun, especially when students have time to "tinker." Some experts believe that tinkering can lead to "deep, firsthand learning about scientific concepts" as students formulate ꓯuestions, assess the evidence, then generate valid responses.

In my experience "experiential learning" has the most impact on youth learning about STEM for the 1st time. At It helps to entice imagination and encourage technical development skills to solve problems through team driven initiatives.

2. ENHANCED COMMUNICATION & COLLABORATION

When students enroll in STEM courses in high school, they have recurring opportunities to enhance communication, especially through active listening and formal presentations.

Additionally, students in STEM classes may collaborate on projects; thus, students learn to demonstrate patience, empathy, leadership, and inclusion. As they create lasting relationships, they transform the classroom into a safe, energetic, and interactive learning environment.

3. CONFIDENCE TO ADAPT TO CHANGE

For many teenagers, change can feel unsettling, especially if the individual is trying to acquire a new skill. Still, STEM in high school invites students to take calculated, incremental risks that build confidence.

Meanwhile, these same students often discover the value of setbacks. Instead of equating obstacles with personal failure, they find that complications or delays can build resilience and flexibility. Students develop grit, determination, and courage whenever they embrace the learning process, which includes trial and error.

4. COLLEGE & CAREER READINESS

Once students realize that STEM in high school means STEM for everyone, they can augment their education with unconventional classes that ignite new ideas.

Subsequently, STEM teachers may assign projects requiring self-paced work, thus preparing students for independent tasks in college and careers.

Types of STEM Programs for High School Students

According to projections, STEM jobs will continue to expand; by 2027, computing, engineering, and advanced manufacturing will dominate the market. Moreover, STEM-related companies lead the nation in well-paying jobs.

Consequently, motivated students who register for STEM in high school could land exciting jobs with trailblazing companies and organizations. Consider the following STEM programs currently available to high school students.

Science: A detailed study of the natural world, with its complexities and intricacies, can stir the imagination, and in many instances, students who enjoy STEM in high school end up pursuing science-based careers that help shape society.

Technology: From Alexa to Xbox, technology has changed dramatically over the past ten years; smart technology is shaping the next generation of phones, thermostats, refrigerators, televisions, speakers, and watches. Indeed, emerging technology in every field continues to impact how people study, conduct business, travel, and socialize.

Mathematics: Every day, people apply mathematics principles, whether balancing a budget, calculating the distance between two locations, purchasing a house, or baking a cake. Math is also a basic building block for other STEM fields.

The Importance of STEM for All Students

Many graduates choose STEM majors in college, including medicine, computer programming, web design, app development, engineering, and systems analysis because of the vast career opportunities. By contrast, other students gravitate to business, education, journalism, or the arts. Nevertheless, regardless of career path, everyone can benefit from a STEM background.

Emerging jobs here and abroad depend on STEM skills, and high school students should seize the opportunity to develop competency in science, technology, engineering, and mathematics. It is becoming increasingly more popular each year as technology advances.

CHAPTER 1
DEVELOPING SKILLS

STEM (Science, Technology, Engineering, and Math) skills are the real-world problem-solving skills you can develop in school. You might know that STEM stands for Science, Technology, Engineering, and Mathematics. But what does it mean when we talk about skills related to STEM?

When we talk about STEM skills, we're talking about the individual skills needed to do science, mathematics, and engineering and those needed to use technology effectively.

For example, in science, you learn ways to increase or reduce friction on a surface to make objects move across that surface faster or slower. You use technology to simulate experiments that might involve things that are too large or too small to see easily or that are too dangerous to do in a classroom. Engineering principles are used when you design, build, and test models, like the ones needed to understand how energy is transferred. Finally, you use math skills to analyze and draw conclusions from experiments. More specifically, below are some of the essential STEM skills you'll want to develop and enhance:

Problem Solving

STEM problems re□uire you to quickly make sense of problems as they are presented and work productively to propose real and appropriate solutions.

Creativity

STEM requires hands-on, active participation to solve problems effectively. Students are the drivers of solutions and should ask □uestions, propose ideas, generate and test solutions, and make decisions based on data to understand how to refine ideas further.

Math & Science Skills

The mathematics and science skills you are learning in school are the foundation of STEM and must be applied in pursuit of solutions. The math and science lessons will provide the foundation for evaluating and solving problems. They are both tools to establish rational and justification for solutions.

Engineering-Design Thinking

In solving STEM problems, the use of engineering-design thinking is vital. This stategic process helps you to identify the problem, research potential solutions, build prototypes, test, redesign, test again, and iterate further as needed. Each step in the process moves you closer to creating a functional solution.

Critical Thinking

Effective STEM learning requires you to analyze information, evaluate designs, reflect on your thinking, synthesize new ideas, and propose creative solutions. All of these skills are vital to becoming an independent, critical thinker.

Collaboration

Individuals rarely solve big challenges. Working on STEM problems also involves learning to work as a productive part of a collaborative team. Most STEM professionals work in a diverse team to achieve business goals. Its an effective way for companies to ulitize talent and proficiencies.

Ways To Develop Your Stem Skills

Want to improve your STEM skills? There are opportunities around you, both as part of a collaborative group and as an individual.

Here are some ideas for ways to build your STEM skills through working with others:

Join a STEM-focused club or program in your school or

community.

Organizations like FIRST, Project Lead the Way, and SkillsUSA offer opportunities to build STEM skills through practical problem-solving experiences while also allowing you to connect and learn with other students who share your interests.

Find a local organization or committee focused on solving a problem in your area, and ask to be a member.

Your school district, city or town, or even local youth organizations always have problems they are working to solve, one step at a time. Join a team solving a STEM-based issue—like protecting habitat or improving recycling rates—to put your STEM skills to use. You can learn from experts or other community members and develop an understanding of the applications of STEM all around you.

If you are looking for ways you can build STEM skills on your own, here are some ideas on how to take advantage of the STEM that's all around you in everyday life:

Ask questions!

If you wonder how or why something works, generate a ☐uestion you'd like to answer. Then use your research skills to figure out what's happening and why, or design an experiment to test ways to answer your question.

Focus on the impacts of each area of STEM in your day-to-day life. Write down all the math or science you use in one day. (It's more than you might think!) Consider how you use technology or how the objects and structures around been engineered for efficiency and safety. Noticing the STEM around you will help you identify how we use it and why it's so important.

Use what you know abou to be more efficient in your day-to-day tasks. Think about tasks you or your family do every day. Now think about how you might use STEM to make those tasks easier, faster, safer, or more effective. Trying to find new ways to be more efficient, no matter the task, is an important part of STEM thinking.

Use what you know abou to be more efficient in your day-to-day tasks.
Think about tasks you or your family do every day. Now think about how you might use STEM to make those tasks easier, faster, safer, or more effective. Trying to find new ways to be more efficient, no matter the task
Use volunteer opportunities and school organizations to learn more about STEM topics in addition to the class curriculum; 4-H, Future Farmers of America, DECA, National Society of Black Engineers (NSBE), SHPE, Girl Scouts, and Cub scouts, to name a few.
Ask to volunteer with Community Service organizations or churches

Utilize the GIE Test Prep Workshop Independent Study Program. Pre-employment testing is a common requirement for many energy companies. GIE Test Prep prepares you for the topics often covered on those tests.

Where are STEM skills needed?

Not sure what you will do with all the STEM skills you're building? Here is a look at how STEM skills can turn into exciting careers at all levels and areas of the energy industry.

Check out these examples of key STEM skills and the amazing things they can help you achieve in the real world.

Science skills, like observation and experimentation:
Research scientists focus on asking questions like, "How can we generate energy more efficiently?" They then create investigations to look at ways to improve how we create energy and more. Power plant operators manage the operation of power plants by observing all of the aspects of the plant and then making adjustments to ensure everything is running smoothly, as well as to optimize energy use and generation. Plant technicians use observations about systems within the plant to ensure that all systems operate efficiently and effectively. Technicians run tests or experiments to ensure that the systems are in working order and that any new components behave as expected.

Math skills, like analyzing data or using logical reasoning:

Energy analysts use trends in data related to the energy use of consumers to predict how much energy we'll need in the short-term and long-term future. Technicians in the field solve problems like power outages by analyzing the data presented to them by their tools and then following a logical progression to identify issues and use the more effective method to fix them. Meter technicians take careful measurements from various meters that measure every aspect of the power grid. They then collect, analyze, and respond to the data to ensure that every aspect of the grid works effectively.

Technology skills, like using databases or digital communication tools:

Network architects build systems that allow information about when and how energy is needed to reach all parts of the energy grid. Information networks carry the data that analysts use to assess and that engineers need to improve our energy grid. Lineworkers, technicians, and others working in the field use technology to pinpoint system issues and make appropriate repairs and improvements.

Engineering skills:

Engineers of all varieties, from those specializing in fieldwork or electricity generation to those designing, building, and maintaining power plants, all use the engineering design process to help them identify how to grow and improve our energy grid. Engineers, operators, and even researchers look at parts of our energy system and ask questions about how to make it more effective. They create and test new innovations that will change how we distribute and use energy in the future. Operator mechanics use their engineering-based mechanical skills to understand and oversee areas of a plant or grid that they operate daily. This allows them to solve problems related to their area quickly and effectively.

Stem Careers Include

You've already been there and done that with primary and secondary school. Now you're trying to really commit to what you want to be when you grow up. Well, if you're wanting to start getting into specifics, the majority of careers touch upon the STEM industry. Some may fall into one particular part of it, but many combine all forms and functions.

Healthcare is the main industry that is heavily debated as a STEM career. While nursing & medicine encompass much of the STEM career format, some researchers believe it should be its own individual field since it can be so specialized.

Let's examine other popular, and quite lucrative, STE(A)M careers. Some of the following careers are available as online programs too.

❖ Science Stem Careers

- **Medical Scientists:** As a medical scientist, your main duty is to conduct research to improve the overall health of humanity. Clinical trials and investigative methods are used to research their findings.
- **Chemists:** Busy yourself with atomic and molecular levels and how substances interact with each other on a daily basis. You'll need at least a bachelor's degree in chemistry or a related field.
- **Environmental Science:** It has become increasingly crucial to protect our environment over recent decades. As an environmental scientist, you will use your knowledge of science and nature to discover ways to better preserve the health of our planet.
- **Agricultural and Food Science:** You'll be working full-time to improve the efficiency and standards of the food industry and the products associated with it. Most going into this field get advanced degrees, however, a bachelor's degree may get you in the door.
- **Astronomer:** You'll be looking to the skies, studying stars, planets, and other facets of our heavens. Telescopes, both on the ground and in the skies, will be part of your equipment. You will be part of a research team made up of other astronomers, physicists, engineers, and more STEM-related professionals.
- **Bioinformatics Science:** You'll use a mashup of information technology and computer science in your study

of biology. You may also be designing your own computer technology to help you with your research.

❖ Technology Stem Careers

- **Computer and Information Systems Manager:** Also known as information technology managers and IT project managers, you'll be doing the planning, coordinating, and directing of all computer-related activities wherever you work. You may need a bachelor's degree.
- **Computer Systems Analyst:** You will study a company's computer programming, assess it, and help to design and implement one better suited for its needs. To work in this field, you'll need a bachelor's degree. To learn more, read about what a computer systems analyst does.
- **Web Developer:** (designer and developer are not the same) An associate degree is the minimum requirement for web developers, although there are many who are successfully self-taught. You need to know coding and graphic design because you're in charge of the look and feel of a website, from the front end to the back.
- **Information Security Analyst:** You protect companies and other organizations from cyberattacks by analyzing, planning, and carrying out preventative security measures. As the number of attacks via the cyber-land increase, you'll find yourself more in demand. Bachelor's degrees are the

minimum re quirement to work as an information security analyst.

- **Software Developer:** You're the brains behind computer programs. You can focus on application development or systems software. Or, you could do both! Bachelor's degrees in computer science are the typical route.
- **Computer Hardware Engineer**: Problems happen. Especially when it comes to technology, computer hardware, and software. You'll be fixing software and computer systems, and helping to advance technology through your research, design, development, and testing. To work in this career, you'll need to have a bachelor's degree from an accredited program.

❖ **Engineering Stem Careers**

- **Petroleum Engineer:** Petroleum engineers design and build new oil and gas extraction technology with a focus on protecting our environment.
- **Nuclear Engineer:** You'll use nuclear energy for good, from medical-related to harnessing its power safely and efficiently. A bachelor's degree is a minimal re quirement, along with some work experience.
- **Mechanical Engineer:** You'll need to have a bachelor's degree and a state-specific license to work as a mechanical engineer. It's a pretty broad career; you'll be touching

almost every aspect of equipment for almost every part of life.

- **Computer Engineer**: This is the path most traveled for students wanting to focus on computer hardware engineering. You'll build, test, and analyze computer hardware, including your own creations.
- **Biomedical Engineer:** Medicine and biology come together—you'll create the equipment, software, device, and computer programs used in healthcare. You'll need a bachelor's degree from an accredited program.
- **Marine Engineer:** Naval ships from sailboats to tankers are designed by marine engineers. You'll need to get a bachelor's degree in marine engineering, marine systems engineering, or marine engineering technology. Internships are encouraged since most employers prefer candidates with some experience.
- **Chemical Engineer:** Chemical engineers usually hold a bachelor's degree, and internships are encouraged. You'll be using a little bit of chemistry, shaken with a bit of biology, physics, and math.

❖ Math Stem Careers

- **Actuary:** A bachelor's degree along with the proper certifications will get your foot in the door to work as an actuary. You'll use your background in math, statistics, and financial theory to determine the risks a company could face, and help them to develop policies to minimize those potentials.

- **Mathematician and Statistician**: Your love of all things mathematical can translate to a high-paying career. You will be working with data samples and statistical software to analyze information for research studies.
- **Personal Finance Advisor:** You work with people to help them make wise investment choices; whether it's for a mortgage, estate planning, college, or any other type of investment a person may need to make. Minimally, you'll need a bachelor's degree in finance and the proper certifications. However, this is a very competitive field, so a master's degree in finance may help you stand out in the job market.

There are so many different STEM careers available beyond what was listed here. One thing is certain, STEM fields are among some of the most interesting and lucrative job choices. And if you don't want to sign up for a 4 year STEM program, that's ok. You can probably find some 2 year STEM training programs at your local trade school or technical college.

7 Well-Paid STEM Careers List

STEM careers have different points of entry. Some of these jobs are attainable with an associate degree, while others may require a Ph.D. Regardless of where a STEM career begins, there are well-paid job opportunities available. Here is a STEM career list with 7 of the most lucrative career options:

1. Forensic science technician

National average salary: $50,050 per year

Primary duties: Forensic science technicians analyze crime scene data and evidence. In this role, they work closely with criminal investigators. Besides working with criminal investigators, forensic science technicians spend most of their time in laboratories analyzing different sets of data and providing explanations that ultimately help investigators.

2. Microbiologist

National average salary: $65,533 per year

Primary duties: A microbiologist studies viruses, parasites, and fungi in labs. Their roles typically revolve around researching these elements and providing data on their findings. A STEM career as a microbiologist begins with a bachelor's degree in microbiology or another closely related field. Microbiologists who want to conduct more advanced research need a Ph.D.

3. Cytotechnologist

National average salary: $70,000 per year

Primary duties: Cytotechnologists are medical professionals responsible for analyzing cellular data and bodily fluids. The purpose of this is to highlight and document any anomalies that might exist in the samples they analyze. Cytotechnologists often support other medical professionals, and their work helps adequately diagnose patients who are at risk of cancer and other cellular anomalies.

4. Radiation therapist

National average salary: $71,964 per year

Primary duties: Radiation therapists work primarily with cancer patients administering their radiation treatments. They operate the machines to direct the treatment in the affected area. Besides requiring either an associate degree or a bachelor's degree, most radiation therapists need a license to work in their respective states.

5. Geotechnical engineer

National average salary: $81,611 per year

Primary duties: Geotechnical engineers study and research earth materials. Their research is used to determine the best possible way to build lasting infrastructure. Essentially, they make sure that any kind of construction project has a solid foundation to work on.

6. System administrator

National average salary: $84,619 per year

Primary duties: A system administrator keeps computer networks running. Major corporations hire them to handle large-scale computer networks in charge of managing their operations. The systems administrator oversees the network maintenance and resolves any network issues that arise. Systems administrators who work for companies that use smaller networks provide more in-depth services.

7. Civil engineer

National average salary: $88,149 per year

Primary duties: Civil engineers build and manage infrastructure. They work in tandem with other professionals, like geotechnical engineers, to make sure buildings, bridges, roadways, and other types of infrastructure align with environmental factors and safety regulations. Civil engineers usually have a bachelor's degree in civil engineering, but they may need to obtain a master's degree to pursue civil engineering specialties or senior positions.

CHAPTER 2
ASSESSING YOUR STRENGTHS

As a student, knowing yourself, including the things you are best at and the skills you could do with a little more work, can be invaluable. Indeed, understanding your strengths and weaknesses can help you improve your performance and build on existing success. This knowledge can also demonstrate the critical insight teachers, professors, and future employers will be looking for. With that in mind, why not read this section on the students' most common strengths and weaknesses below to see if any resonate with you?

What Are Student Strengths?

Today's students are often labeled as being either smart or not. The truth is that we all have strengths and weaknesses. Our strengths may be different from others, but they are still there.

A student's strength can be any personal trait that stands out about a student and helps them do better in school, whether they're able to memorize information quickly or have good stamina for mental activity.

It is time to stop focusing on the negative things about ourselves and instead find our true selves by taking a closer look at what makes us who we are. We need to take pride in our uniqueness because it makes us special. Strengths are unique aspects of a student's character.

Talking about student qualities doesn't seem like something that would generate much attention, but it should.

Thousands of students go through school without hearing positive feedback about themselves or their work from an instructor or teacher, making them feel like they don't matter and everything they do is not good enough because nobody tells them differently.

If you know your student's strengths, you will always know how to make up for your weaknesses when faced with challenges during your student life.

Once you start paying attention to what you do best, a pattern will emerge, and success will follow.

Student strengths should affect not only academic work but also emotional health.

There are some qualities students might need more than others, while other student qualities may never come into play, so don't feel like you just missed something when this happens.

List of Valuable Student Strengths

List of most common student strengths:
- energetic
- kind
- creative
- adventurous
- trustworthy
- playful

Student Strengths – 4 Examples Fully Cover

Evaluating Yourself Sincerely

The first step to making the most of your education is being honest about your performance in school.

It's not just about good grades but every aspect of what you do as a student and person.

Take time for self-reflection to get an accurate picture of where you stand academically and mentally so that you can start taking steps toward improvement.

This means that you should be proud of your accomplishments but also willing to admit when you're not up to par.

Do not worry if it is hard for you; instead, take that first step towards evaluation because it can make all the difference in your student's success.

Being Open With Those Your Trust

One thing that helps (but that most students find difficult) is talking with someone you trust about what's going on in your life.

You might be surprised at how helpful just one conversation can be.

Sometimes students can feel alone in the world, but that's not always true.

By talking to someone who can give you unbiased, non-judgmental advice and support, you can find yourself going to the next level as a student without much difficulty.

Other resources available for student success may not exist on campus, such as counselling centers, friends' houses, extended family members' homes, etc.

Creates Self-Awareness and Motivates Improvements

It's not enough to go through the motions.

You have to know that you're going places and give yourself a chance to get there. That's why it's important to keep track of where you stand and the things to work on.

Don't worry; it's not that difficult.

All you've got to do is take some time out of your day (maybe in the morning before class or at night after dinner) and jot down three things that went well today and how they made you feel great about yourself.

This could be anything from getting an A on a test to receiving praise for doing something right by someone important in your life.
You might also want to add any other goals or achievements you may have achieved, too, as this can inspire when times get tough.

Builds Confidence in Themselves and Others

There are many ways that a student becomes more confident, but sometimes it's up to the students themselves to take charge and determine what will work best for them.
One way of doing this would be by setting aside time each day to reflect on your strengths, such as goals or achievements that make you feel good about yourself.

Another suggestion is to be mindful of any compliments people have given when they say nice things about you- it's always a great feeling being appreciated.

If these don't seem enough, try one more thing before going with another idea: ask a trusted friend, parent, or tutor what they see as your student's strengths. Then, help others become more confident by doing the same thing with them.

List of Most Common Student Weaknesses

The most common student weaknesses include:
- Stubborn
- Lack of focus
- Fear of failure
- Procrastination
- Dramatic
- Inactive

What Are Student Weaknesses?

Student weaknesses prevent students from performing at optimal levels for success. It is an advantage to know ones natural gifts and talents. Emphasizing your strengths create a fast tract to seeing positive results, which can be a wonderful energy and moral booster.

It is also helpful to understand weaknesses to manage effective solutions to improve them.

Memorization

Let's have a close look at some of these weaknesses. Many students find that they are not good at memorizing facts. This means they don't know the information they need to recall for their work.

Some students are very good at memorizing facts but have a hard time understanding what they're being taught in class- it just goes over their heads and makes them frustrated.

No student can do everything well; we all have strengths and weaknesses when learning, so be encouraged because no one has it all.

Learning Styles

Understanding your prefered learning style will tremendously help when working to address weaknesses. Most people that have problems processing information may not have figured out their preferrred learning style.

There are 4 types of learning styles, visual, auditory, kinesthetic and reading/writing learners. Knowing your learning style provides insight into the best way YOU should receive new information. For example, coupling photos or pictures with a history lesson may enhances memorization of significant dates and times.

Health Deficiencies

A student's weakness could be due to lack of sleep or nutrition, poor memory skills, or anxiety about upcoming tests.

How to Apply & Utilize Your Strengths

A student's weakness is not always bad, and there are always ways around it.

For example, if they're having difficulty retaining information from the last time they studied for a test but excel at memorizing facts – maybe they should try to take notes during lectures to remember better what was said.

The student who has trouble understanding the the material could ask classmates for help before class or after school with homework assignments they understandnd what material is being taught.

Awareness of these weaknesses allows us to find solutions that work best for each student rather than relying on one-size-fits-all approaches.

Be Humble

The first step in being humble as a student is being grateful for what you have and understanding that everything takes time. You are not alone in your struggles and failures; everyone has them at one point or another.

It's important to remember that nobody wants to be told their opinions don't matter because they're just kids, but at the same time, everyone has to go through this at some point or another.

Remembering these things will help you stay humble while you work towards your goals of becoming successful and achieving student success with student strength.

That humility is what makes a truly great student. You can do anything as long as you never stop believing in yourself and showing dedication by practicing each skill over time until it becomes second nature.

The school takes a lot of hard work, patience, and dedication. Using your student strengths (while staying humble) will help you succeed as a student and in life.

Get a Mentor

There are many benefits to having a mentor as a student.
Mentors can guide you in all areas of life: schoolwork, college applications, career aspirations – even emotional support.

Get help from someone who has been there before so you can be successful now and in the future. One of the most important benefits is that mentors can help you find your strengths and teach you how they can be used in college.

A mentor will also advise you on what to do when facing challenges, like making friends or overcoming procrastination. They will also listen and provide feedback on any problems with your schoolwork. The best way to get a mentor is by reaching out. Ask around and see if anybody would be willing to spend some time helping you succeed at school.

Personality Test

Interview your friends and Family and ask them what you are best at.

Anecdote: I am from New Orleans, La, home of the annual Essence Festival. Every year my mother would bring me to see the main stage performers like Prince and Mary J Blige but in my senior year of high school, I was asked to coordinate a team of friends that would assist in selling battery-operated roses that glow in the dark. I set a meeting to discuss the plans with my mother and my friend's parents. I shared with them the benefits of joining the sales team and ensured that all expectations were clearly defined for the sales leader and the team. We had a blast selling our "Frankie Beverly roses" that year. I didn't realize then that the experience would impact my career path later in life as a project engineer, leading multimillion-dollar construction projects with team members of various disciplines, from start to finish.

I developed skills early on that provided insight into my ability to harness the power of people and guide the team to meet a mutual goal.

Tools That Helps Improve Your Strengths

Below are the helpful tools to improve your strength:

- Myers Briggs Type Indicator

- Bennett Mechanical Comprehension Test (BMCT)
- Differential Aptitude Test (DAT)

These tools are comprehensive functional and aptitude tests specially designed to assess the skills and performance of a candidate as per the required industry standards.

These tests are designed to assess what a person is capable of doing or to predict what a person can learn or do given the right education and instruction. They represent a person's level of competency to perform a certain type of task.

Both the last two (Bennett Mechanical Comprehension Test (BMCT) and Differential Aptitude Test (DAT) are for grades 7 to 12. You can stress the value of having a baseline understanding of your level of understanding at each grade level to gauge career interest for the future.

STEM careers can be enriching and fascinating. However, not all students feel they leave high school well-prepared for them. Utilize project-based learning, keep in mind the challenges minority populations face in these fields, and collaborate to create an effective curriculum. A lot is already expected of teachers, but with regular adjustments to practices as a high school student, you can thrive in STEM careers.

CHAPTER 3
NETWORKING

What Is Networking?

Networking has become increasingly important in today's world. It is vital at every career stage and can open doors when entering the workplace. You never know the value of your network until you need to connect with a contact for support or guidance. Ultimately, networking is reaching out to similar-minded people to share ideas, knowledge and possibilities. It can be very collaborative. Networking also enables you to expand your knowledge in your area and offers a flow communication that can be beneficial for both parties involved.

Networking may seem daunting to high school students, with thoughts of meetings and suits arising in a stiff professional setting. The truth is, though, times have changed so dramatically that the best networking often takes place in an informal social setting. The casual and relaxed atmosphere promotes honesty and straightforward communication and, more often than not, creates a new friendship.

Undoubtedly, on some level during their career, students will have more formal opportunities for connecting with peers or leaders. Still, at high school age, networking will rely heavily on exchanging information and ideas, mostly on an informal level. Therefore, to redefine networking for those young, eager, energetic, and intelligent high school students, I would say that networking is:

"The desire and ability to learn something new and share your thoughts through interacting with people you may not know already."

Why Is Networking So Important?

Everyone is told that networking is the key to a successful career, but do you know WHY is it so important?

1. You may connect with a potential mentor through networking events! If you find someone you look up to who agrees to be an informal or formal mentor, they can support you with many things, such as improving your CV, sharing opportunities with you, and helping you break into the world of work.

2. Networking with other people in the field you are passionate about might mean you get to hear about opportunities before other people do. Many jobs, internships, and more are not advertised immediately, and if you know about it before the public, you have a better chance of landing a coveted position!

3. You might find that, through networking, you become a lot less worried about the world of work. A lot of our workplace anxiety is due to the fear of the unknown, but through networking, we learn the ins and outs of our field and see confidence increase as a result!

4. You can leverage your network by being exposed to the best practices for your field of interest, new trends, challenges, and the future of your industry, which you can discuss in interviews to show that you know your craft.

How to Start Networking: Top Tips and Tricks

Starting to network with peers and professionals can feel awkward at first. Here's some advice to get past your initial reservations and go for it.

There's a myth going around that networking is a necessary evil. Someone told college students that it's hard to connect with employers/professors/random strangers working in their desired field. Getting into events is expensive, and even if you get in, there's no guarantee that you'll talk to the right people. Here are a few tips to disprove negative ideas.

Just talk

The most important part of networking is talking. That's all there is to it. Everything else is supplemental and starting a conversation can be easy! If a speaker comes to your lecture, don't miss an opportunity to build a relationship. Please stick around and talk to the speaker: ask them about a specific remark they made, comment on how you agree with their perspective, and ask for an informational interview. Just. Talk. To. Them. It's so easy to connect with anyone around you—all you need to do is is muster up some courage and start talking.

In college, I purposely saught jobs that allowed me to have access to "decision makers". I worked as a server in the Louisiana State University's corporate suites. I knew that I would likely engage with company executives. It was an informal way for them to meet me and develop and rapport, which would allow me to ask questions about opportunities. It worked like a charm. While I served them cocktail shirmp and little cute cheese hors d'oeuvres, I was also serving my resume and credentials for a prospective job. Not only was I able to speak to a department leader, his wife gifted me Chanel sunglasses!

Have your info ready

Once you feel you've made a good connection, you should ideally exchange information. A printed or vitual business card makes that so much easier. Create a business card as soon as possible. You can easily make them on programs like Canva or design and have a few printed on the vistaprint website. It doesn't have to be anything fancy, and it's okay if you don't have a job title yet. All you need is a simple card, one you print out that says your name, your major, and a bit of contact info (i.e., email address, phone number, social media handles, etc.). If you have a website, all the better, but you likely won't need one until at least junior year.

If you're a graphic designer, you should create your design because it's the easiest way to showcase your talent up front. Pro-tip: if you leave the back blank, you can write additional information. For example, if you're recommending a conference, book, or website someone should look into, you can jot it down on the back. Additionally, this connects you to that recommendation, so they may reach out to you later to discuss it.

Follow up

It is important to keep your contact engaged on some frequency once you've met. If you talk to someone you hope to continue networking with, email them within 48 hours to follow up. If you wait too long, you risk them forgetting the conversation, and that's a situation you don't want to be in. It doesn't have to be complicated, just a note that says:

"Dear [His/Her Name], It was a pleasure to meet you at [event name]. I enjoyed talking to you about [something you talked about]."

A follow-up will further establish your connection, and you need to stay on top of it, especially if you talk about any opportunity, like an informational interview or internship. And don't stop there with follow-ups. Continue to connect, by following them professional social media portals such as Linkedin. Also, acknowledge their accomplishments so that you show interest and there is a balanced exchange of connectedness.

Know where to network

Yes, all that's well and good, but where do you network? The answer: Everywhere.

As a work study job, I was able to snag a role in the Chancellor's Oficce. I knew that working in the office I might meet academic leaders and partnering corporate leaders. Many of the executives for the companies like Exxon Mobile, Coca Cola, Shaw Group and Entergy were major donors to programs hosted by the college.

Networking doesn't always require going to a fancy event you paid over $100 to attend. You can network at the grocery store when someone is sporting a tote bag from your school or research free events relative to one of your interests (poetry/Moth readings or events at the public library). Your college may also provide networking opportunities to you, such as alumni events, internship, and job fairs. Check with your school's career services office for events available to you.

Set realistic expectations

Don't go into a networking opportunity assuming you'll walk away with everything you've ever wanted. You could talk to the CEO of a major company and walk away without any new knowledge and a dead connection. Or you could talk to someone working part-time at a local nonprofit and get a lead on an internship or two. You could even talk to someone with no relation to your interests, but they have a friend you should meet and will pass your information along to them. Go into networking opportunities with an open mind and no plan. You'll get more out of these connections if you don't try to force a specific outcome. Finally, don't forget that networking isn't just about furthering your career. You can network strictly to meet interesting people, especially since you might end up being the person they need the connection from, whether it's recommending someone to fill a position you aren't qualified for (but your friend is!) or the person who shows them how to use Twitter successfully. Sometimes all you walk away with is a new friend, and that's just fine.

Networking is indeed necessary, especially as you get further into your college career and start looking for internships, volunteer or research opportunities, and (shout out to the seniors) careers, but it's far from evil. You're talking about things you're interested in with like-minded people. Now that's fun!

Four Easy Networking Strategies for Online Students

Networking as an online student takes a little extra effort than if you were on campus. Here are some simple ways to start connecting with peers and professors

As technology advances, online education is becoming more popular than ever. Online programs and schools create new ways for students to pursue higher education that fits their lifestyle—including students who otherwise may not have access to higher education at all. Online universities are becoming more credible in the workforce, with many employers acknowledging the accreditation of virtual degree programs.

There are many differences between traditional and online degree programs, but one major difference is the in person interaction. In conventional settings, students can talk to each other in class, see a professor after a lecture to ask questions, or head to the advisement office to connect with a mentor. In an online setting, there are still plenty of opportunities for students to network and make connections; they have to approach them a little differently. Let's look at four simple and effective ways online students can foster relationships and build their networks.

1. Utilize school resources

Most online schools offer a range of resources to help students networks, such as LinkedIn and Facebook groups, chat rooms within their student portals, and discussion boards. If you're an online student, you must ask about and utilize these resources to establish connections and create relationships with your fellow students. These resources may also be available for alumni, allowing online students to network and make connections that could lead to relevant job opportunities, even after graduation.

2. Use social media to connect

If you're in an online class or on a school discussion board, chat with other students as much as possible. Then take it a step further by asking to connect with them on social media. Social media is a great place to connect and deepen peer relationships. Connect on LinkedIn, follow them on Twitter, or become Facebook friends. Consider asking for their permission first so as not to seem intrusive. These online schoolmates can be great resources for discussing schoolwork, projects, and job opportunities.

3. Build a rapport with professors and faculty

Online professors are as eager to connect with their students as traditional professors. They typically provide students with email addresses or chat options for a reason, so use them! Reach out to your professors with questions, and work one-on-one with them to make sure you understand all the material correctly.

Many online schools also have advisors or mentors who will work with you and answer all your questions. Connecting with these faculty members via email or even phone chats can help you navigate your college experience more easily. Take advantage of the opportunity and get to know the mentors and professors at your online school—you'll be more successful when you enlist their help.

4. Don't be afraid to start up a conversation

Joining online study groups or sending emails to classmates with whom you don't have personal relationships can be nerve-wracking. But remember that they're students just like you looking to network and make connections too. You have to take the plunge and start the conversation! (This applies to your professors too, who are eager to answer questions and have discussions with you.) Finding the courage to ask questions or start a friendship can transform your education and career path.

As an online student, networking is key to having a positive experience that sets the stage for your future. Your professors and classmates want to network as much as you do, so get up the courage and go for it. You never know how a relationship might change your future.

How Technology Plays a Role in STEM Education

Technology gives students access to STEM education resources. Using mobile devices, students can participate in virtual labs to conduct simulated experiments using university-level equipment. Experiments can be repeated so students can correct and learn from their mistakes.

Students can also use mobile devices like smartphones and tablets to watch STEM programs. STEM-oriented shows geared toward different age groups are available on networks such as PBS, Discovery, and National Geographic.

Special tablets are available that are pre-loaded with STEM-focused content. These STEM tablets can be used to conduct experiments or participate in learning modules. Students can use these tablets individually or as a group to collaborate on a project.

4 POLISH YOUR PROFESSIONALISM

Professionalism is an essential soft skill in today's workplace, and high school teachers are often expected to teach these skills to their students.
You need to prepare students for their future careers, but finding or creating activities to teach professionalism that will resonate with your students can be time-consuming and seemingly impossible!

Soft skills are interpersonal and personality traits that go beyond technical and analytical skills. Of course, having a solid STEM education and being a star student is a must, but to differentiate yourself for a STEM job, you must also show a certain emotional IQ.

Ten Skills Needed to Land a STEM Job

Here are ten skills students need to work on to ensure their future success in STEM-related jobs.

1. Critical thinking. Tech leaders face problems daily, and the course they choose to take in solving them can have far-reaching implications. We look for candidates who have honed their critical thinking skills to make the best possible decisions. Early in their education, students must learn to define and approach problems and situations from many different viewpoints, analyze every possible solution, and anticipate each option's consequences and outcomes before taking action.

2. Analytical skills. It's important for students to develop skills to analyze data sets and to understand how they relate to other data, systems and processes. The ability to synthesize and interpret complex information from multiple sources is a huge indicator of success in our business.

3. Problem-solving. Efficiency and value are paramount for our clients, who want solutions that use the smallest amount of effort to the greatest effect and can solve multiple issues simultaneously. In the interview process, we ask candidates how they would solve a hypothetical problem. This exercise tells us whether they tend to layer on complexity or distil a problem down to its essence. An

ideal job candidate demonstrates the ability to devise the simplest yet most effective solution.

4. Innovation. In an interview, we look for a candidate's willingness to take risks and offer creative, even quirky ideas. Ultimately, a fresh perspective and a spirit of inventiveness will outshine those that follow the same old path.

5. Collaboration. No employee is an island, so job seekers must develop the skills they need to be team members who can work with others toward a shared goal. Recent college graduates may not have extensive work experience to demonstrate this ability. Still, they can share a story about a time when they were part of a high-performing team or group at school or elsewhere. In an interview, talk about how you contributed directly to your team's success, how you learned from the challenges of working with others and how your working style has evolved as a result. This information helps recruiters understand how well you will relate to and collaborate with others in the workplace.

6. Communication. You may be a rock star in your technical field, but you'll be at a major disadvantage a job seeker and employee if you aren't able to communicate your ideas. Students should learn to convey complex ideas to people of varied backgrounds and job titles, including those who have less or more technical expertise than they do.

7. Customer orientation. The most successful businesses make the customer's needs priority number one. Our employees are invested in creating a meaningful experience for each customer while designing solutions that help them achieve their goals today and ten years from now. They use the skills we encourage STEM graduates to focus on: active listening to understand the clients needs, and going above and beyond the call of duty in customer service.

8. Adaptability. It's almost impossible to stay up to the minute with all the skills and systems you'll need to be effective as you move from college to career or when you land a job at a new company. You can show how you have gained knowledge and new abilities ⬜uickly in past positions and offer evidence of continual career development, such as recent education, certifications, promotions and training courses.

9. Social responsibility. We look for candidates who share our values and can demonstrate consistency with those values in their decisions, actions, and how they work with others.

10. Balance. Employees who have a passion outside of the workplace are more productive and more satisfied with their work and physically healthy. We love it when job applicants use these activities to demonstrate ⬜ualities they are looking for, such as

teamwork, leadership or perseverance.

Famous artist, Salvador Dali once said, "Intelligence without ambition is a bird without wings." All the knowledge and technical training in the world will get you nowhere unless you have optimism and ambition driving you every day. These powerful forces are contagious in the workplace.

Select Professional Organizations For Math And Science Students

Students are encouraged to join professional organizations of their major discipline. Here are some suggested organizations to explore:

American Chemical Society (ACS)

ACS student chapters are organizations for undergraduate chemistry majors. Members participate in various programs and activities that enhance their college experience and prepare them for successful careers. Dues for ACS members are $25 per year for undergraduate students and $55 per year for graduate students.

American Mathematical Society (AMS)

Students at all levels can benefit from AMS programs and services. In addition to offering various programs and opportunities for students, the AMS collects information about opportunities of interest to students in the mathematical sciences. Membership for undergraduate or graduate students costs $52 per year and covers career and professional development services provided through AMS.

Mathematical Association of America (MAA)

MAA membership is your portal to the community of mathematicians and mathematical educators, all focused on advancing mathematics at the collegiate level and providing you with opportunities to build your network of professionals and peers. Student membership costs $35 per year and includes full access to online subscriptions to all MAA journals, member discounts and much more.

American Society for Microbiology (ASM)

The American Society for Microbiology is the world's oldest and largest single life science membership organization. The American Society for Microbiology aims to advance the microbiological sciences as a vehicle for understanding life processes and to apply and communicate this knowledge for improving health and environmental and economic well-being worldwide. ASM supports programs of education, training and public information; promotes the contributions and promise of the microbiological sciences; recognizes achievement and distinction among its practitioners, and sets standards of ethical and professional behaviour. Student membership is $35 per year.

Federation of American Societies for Experimental Biology (FASEB)

FASEB's members are scientific societies that share a common vision for advancing research and education in biological and biomedical sciences. There are 27 constituent societies, and you can find more information about them on the FASEB website.

American Institute of Biological Sciences (AIBS)

The American Institute of Biological Sciences was formed to bring together the organizations and individuals that advance the biological sciences to work together on matters best addressed through united action. The AIBS is an organization that informs and leads research, education and policy-making at the frontiers of the life sciences. AIBS listens, anticipates, advises, collaborates and when needed, leads efforts in the life sciences community to address scientific and societal challenges.

American Society for Cell Biology (ASCB)

ASCB is an inclusive, international community of biologists studying the cell, the fundamental unit of life. ASCB is dedicated to advancing scientific discovery, advocating sound research policies, improving education, promoting professional development and increasing diversity in the scientific workforce.

Membership in the ASCB is open to all research scientists, students, educators (high school, undergraduate and graduate level) and technicians who have education or research experience in cell biology or an allied field. Student membership is $26 per year.

Society of Physics Students (SPS)

The Society of Physics Students is a professional association explicitly designed for students. Membership is open to anyone interested in physics. Besides physics majors, members of SPS include majors in chemistry, computer science, earth science, mathematics, and other fields. Membership dues for this society are $24 per year.

American Physical Society (APS)

APS advocates for research funding and science education, provides scientific expertise on issues, such as energy and the environment, and informs the general public about physics. APS members benefit from many educational and diversity programs. Students enrolled full-time in an educational institution ualify for membership at the reduced rate of $25. And as a student, your first year of membership in APS is free.

IEEE Computer Society

When you join the IEEE Computer Society as a student member, you become part of a vibrant technical community. You have opportunities to connect with people who share your interests. Student membership costs $20 per year. As a student member, you receive free, unlimited access to the Computer Society Digital Library—the authoritative source for everything in the computing universe. In addition, student members get free Microsoft software, a $2,000 value.

Association for Computing Machinery (ACM)

ACM, the world's largest educational and scientific computing society, delivers resources that advance computing as a science and a profession. ACM provides the computing field's premier Digital Library and serves its members and the computing profession with leading-edge publications, conferences and career resources. Student membership costs $19 per year.

American Geosciences Institute (AGI)

AGI is a federation of geoscience societies which provides information and educational services to its members, promoting a united voice for the geoscience community. AGI consists of several member societies listed on the AGI website, which provides more detailed information about each society.

The American Association for the Advancement of Science (AAAS)

The American Association for the Advancement of Science (AAAS) is an international non-profit organization dedicated to advancing science worldwide by serving as an educator, leader, spokesperson and professional association. In addition to organizing membership activities, AAAS publishes the journal Science, and many scientific newsletters, books and reports, and spearheads programs that raise the bar of understanding for science worldwide.

Society for the Advancement of Chicanos and Native Americans in Science (SACNAS)

SACNAS is a society of scientists dedicated to fostering the success of Hispanic/Chicano and Native American scientists—from college students to professionals—to attain advanced degrees, careers and positions of leadership in science. The basic one-year membership for students costs $10 per year.

Association for Women in Science (AWIS)

AWIS membership is open to all women and men who support equality for women in science, technology, engineering and mathematics (STEM). An AWIS membership offers women in STEM a national and local platform to tap into the power of an already established community of women across all disciplines and employment sectors. The student membership is $35 per year.

Association for Women in Mathematics (AWM)

The Association for Women in Mathematics aims to encourage women and girls to study and have active careers in the mathematical sciences, and to promote equal opportunity and treatment of women and girls in the mathematical sciences. Student membership costs $20 per year.

Animal Behavior Society (ABS)

The ABS is a society targeted to individuals interested in forming a network of other animal behaviourists through research and conferences. Student membership costs $25 per year, which includes online access to the leading journal Animal Behavior, an international journal that contains research papers, commentaries, review articles and essays.

Association of Environmental and Engineering Geologists (AEG)

The Association of Environmental and Engineering Geologists contributes to its members' professional success and the public welfare by providing leadership, advocacy and applied research in environmental and engineering geology. AEG is the acknowledged international leader in environmental and engineering geology and is greatly respected for its stewardship of the profession. AEG offers information on environmental and engineering geology, which is useful to practitioners, scientists, students and the public.

Other geosciences organizations recognize the value of using and sharing AEG's outstanding resources. Membership in this society for full-time students is free.

National Science Teaching Association (NSTA)

NSTA is an association that provides resources and professional development for students and professionals in the field of science education. The digital discounted student membership is $40 per year.

National Council of Teachers of Mathematics (NCTM)

NCTM is an association that provides resources and professional development for students and professionals in the field of mathematics education. The student membership costs $49 and includes digital journal access, complimentary regional meeting registration, and a discount at the NCTM store.

With a great education and technical skillset, create a mindset that shows thoughtfulness and perspective to make yourself the ideal candidate for a STEM career.

Ten Reasons To Consider A Professional Career in the STEM Fields

Thinking about pursuing a career in science, technology, engineering, or mathematics? Now is a perfect time. STEM is a quickly growing and high-paying area that is an excellent direction for new or graduating students to take. But is it right for you? Check out these ten reasons why you should consider making STEM your future career.

1). Cross-Marketability of Skills

It's not like if you're a biologist, then you can suddenly go into theoretical physics, but the basic skills you learn are useful in many different professions. By studying to follow a field in science or technology, you'll be learning math, research methods, problem-solving, and how to fill out mountains of paperwork. Many jobs across all fields need that, which means you're already qualified. If you decide to change jobs or have trouble finding one right away, you won't be left out in the cold.

2). High Pay

Going into STEM isn't just a great way to get a job. It's also a way to get a high-paying job. Studies have shown that 63 per cent of people with a degree in STEM-related work get paid more than someone with a bachelor's degree in anything else. More than that, 47 per cent of people with a bachelor's degree in a STEM field make more than people with a PhD in other areas. You'll be more likely to make lots of money than someone who majors in another field of study.

3). Job Availability

With this economy, you might be worried about how you will get a job. Are there that many jobs available to you? The short answer is yes. There are STEM jobs in numerous different fields ranging from research assistant to physics educator. More jobs are always added, so you're never going to find a scarcity of positions for someone with your qualifications.

4). Less Competition

Even though there are plenty of jobs, there are still probably hundreds of people clamouring to get at them, right? Wrong. Every year 3.2 million jobs in the STEM fields go unfilled, mostly because no one is qualified to fill them. By educating yourself and preparing to go into one of those jobs, you'll be a head above the rest who don't have what it takes to qualify.

5). Less of a Gender Gap

Ok, so maybe you're a woman worried about that huge pay gap determined by gender. This is yet another way STEM can help with your life. Studies have shown a much smaller salary gap between men and women in the STEM fields. It's also got plenty of racial and gender diversity, so if you're looking for equality in the workplace, then look no further.

6). Basic Skills Won't Get Obsolete

Indeed, technology is ever progressing, but the basics stay the same. The world will still need math and still need research. Because you have a basic knowledge of the scientific method, computers, essays or report writing, and much more, you'll be valuable no matter how things innovate. You'll be more ready to adapt to the changes than people without experience in the field.

7). Innovation

All that innovation doesn't just help you get a nice paycheck. It also helps others. You can work in cutting edge of fields like medicine, computer technology, robotics, and more. If you have a humanitarian streak, what better way to show it than by gifting the world with your innovative ability?

8). Better Classes

Right now, the government is making a big push to get classes funded and available for STEM students. There are more varied classes in different areas, and most states' funding is better for STEM-related classes. Given that half of the high school students say they're not going to go into STEM majors, you may also find that your classes are smaller, which gives you better access to professors and resources.

9). Everyday Critical Thinking

Being a problem solver in the classroom and at work can lead to you being a problem solver in life. Training in a STEM field can assist you in your everyday world and even save you money. If you have a better grasp of computer engineering, you may not have to take your computer to the IT guys so often. If you're a math wiz, doing your taxes might not be as big a bother. Besides, who else do your friends know that can make a potato cannon out of office supplies or a fighting robot from car parts?

10). Love

There is no better reason to go into a field than this. For some people, science, math, and technology are not just jobs; they're a passion. If your career is something you enjoy doing, you'll lead a happier and more productive life. You'll be one of the few who doesn't dread going to work every day and can brag about your job. Even if your grades weren't the best and your pay isn't as high as you'd like, pursuing what you love is a reward.

5 PREPARE FOR GREATNESS

What is Greatness?

Someone with greatness distinguishes themselves from others, sets themselves apart, and is truly eminent in their field.

Or as it is defined:

Greatness: The quality of being great, distinguished, or eminent.

You may be the type of person who takes that inspiration and runs with it. Or you may be the type of person who looks on admiringly, assuming you'll never be like that. Ask yourself:

- Are you a leader?
- Are you a follower?

Here's the thing: You do NOT have to be a leader to be great. Greatness is essential for both leaders and followers.

How can you achieve greatness with everything you do? Can you be a great follower? A great leader?

What if there was a formula to create greatness . . . in anyone?

A Story of Greatness:

That is the question Lewis Howes has asked himself for most of his life. As a kid, he was the rejected, awkward student who struggled in school and had no friends. But once he discovered his passion and talent for sports, he began to channel all of his energy and frustration into athletics.

That led to many achievements in football and the decathlon (including breaking a world record), but it wasn't enough for him. He still wanted to be great.

After a career-ending injury, he found himself recovering on his sister's couch, broke and clueless about what to do next. But he again channeled his frustration and energy into learning new skills and built a multi-million dollar business as an online marketing consultant in just a few short years. But it wasn't enough for him. He still wanted to be great.

He sold half of the company to his business partner and returned to the drawing board. This time, he started to think beyond himself. What would make a truly great impact on the world? He realized he needed to be focused on serving other people.

So he started a new kind of school called The School of Greatness. In the form of a podcast, he began to interview the best and brightest minds in the world about their insight on what makes greatness.

Today, he has published hundreds of episodes, has millions of downloads, and shares the microphone with superstars like Tony Robbins, Julianne Hough, Arianna Huffington, and Scooter Braun.

Preparing For Greatness In STEM

School is such an important formative experience for everyone. It helps guide us toward our interests, develops our social foundations, and shapes our understanding of who we are and who we want to become. For some students, this is when they develop their interest and enthusiasm for science, technology, engineering, and math (STEM) fields.

However, as they reach high school, there is often a shift in educational intentions. The responsibilities of teachers shift from simply providing insights into these subjects to actively helping to put students into a position where they can prepare for a STEM career. This isn't always easy, particularly in schools where inadequate funding results in less access to STEM materials and fewer opportunities to engage meaningfully with the subjects.

Science, technology, engineering, and math (STEM) skills are important for career success.

You might think you're in a career that doesn't need STEM skills. But you'll find that a little STEM knowledge will help you succeed in any career.

It takes persistence to do college-level work in math, science, and technology, but the rewards can be very high. STEM Careers are in demand and tend to pay very well.

You can gain STEM skills through academic courses, work experience, and hobbies.

STEM Classes

Get started by choosing these classes in high school or college.

- Algebra
- Statistics
- Biology
- Geography
- Physics
- Chemistry
- Computer technology
- Computer-assisted art

- Research methods (in any discipline)
- Calculus
- Fluids
- Electronics
- Dynamics
- Political science
- Technical writing (business plans, research reports, grant proposals, etc.)

STEM Experience and Hobbies

Try one or more of these activities to increase your STEM greatness skills:

- ✓ Using budgeting and math skills, help a community organization with a fundraising event or project.
- ✓ Participate in a data-collection project to gain experience writing surveys, conducting interviews, and analyzing data using standard database software.
- ✓ Teach youth at a science summer camp or after-school program.
- ✓ Try programming your own video game or customizing your favorite. You can often download free software from the Internet.

- ✓ Explore a technical hobby. Learn about computer parts, try building your computer, and use an online forum to share how you did it.
- ✓ Job shadow or intern with an engineer or someone in another technical field.
- ✓ Join a Math or Science Club.
- ✓ Prepare a project for a science fair.

Learn computer applications or technology related to a field of interest, like CAD for architecture or SQL for database work.

What Makes a Great STEM Student?

Do you currently struggle with science and math or generally find STEM fields disengaging? If so, this does not mean you're not equipped to do well in these subjects in college. It's more likely you have what it takes to be a scientist or engineer if you didn't succeed in these subjects in high school. While educators and the world of science lament the fact that they don't have enough people to fill STEM jobs, many potential scientists and engineers have sworn off these subjects because of their high school experiences, whether it was getting a bad grade, disliking a teacher, or struggling with the material. People with mathematical and scientific talent often pursue professions where they won't feel fulfilled or will never have an opportunity to showcase their genius just because of these past experiences. But you can be a great STEM student.

5 Ways to Become One!

Here are five ways to give these subjects a second chance.

1. Discover what makes a great STEM student

After decades of studying the attributes of the most successful scientists, engineers, and mathematicians, I've identified the skills, traits, and attributes that many of them share. I've also spent time researching the attributes required to be successful in STEM classes from kindergarten through 12th grade. I've discovered that what it takes to be successful in STEM is drastically different from what it takes to be successful in STEM classes. This is a provocative revelation for some and a hidden truth for others. Either way, students must know as they enter college and start deciding what they want to do with the rest of their lives.

Understand what it means to be science-minded

Science-mindedness captures scientific habits of mind, inquiry skills, and observing and learning from the greatest scientists and mathematicians. It includes having or making keen observations, basing claims on evidence, expressing curiosity, being analytical, having profound creativity, being open-minded, and using analogies to make connections between complex ideas. These traits are shared by people who may feel far removed from intellectual pursuits but have much in common with historical geniuses like Albert Einstein and contemporary science heroes like Neil deGrasse Tyson.

Engineer yourself into a scientist

Every day, without realizing it, people who've had negative experiences in school operate with a level of genius that could find them success in STEM fields. The bottom line is everyone can and should pursue these disciplines, even if they've had problems in the past. It's important to acknowledge that whatever reasons you've chosen to disengage from these subjects are justified—but they can be overcome so you can reclaim your STEM genius.

2. Reclaim words said and unsaid

Your environment is a crucial part of your childhood influences, playing a significant part in who you become. If you're in a positive home or school environment, you'll hear phrases that affirm you for being smart and kind. It's not foreign to hear a two-year-old being called a math wizard. But most people aren't properly celebrated for being exceptional. Unfortunately, many students hear words that undermine their intelligence or cause them to see little value in themselves. In addition to the words said, the words that are never used to describe us also determine how we feel about ourselves. You can claim these titles and traits if you choose to, but if no one in your life has called you a scientist or a mathematician or you've never been described as having the science-mindedness traits, you may have determined your apitipude in STEM based on that.

3. Write down your science mantra

To counteract this phenomenon, it's essential to reverse-engineer yourself out of thinking you're not smart or good enough and into who you are and who you can be. Each new frontier on the college journey is a new opportunity to reinvent yourself, and this process begins with one basic step. Write down a "science mantra" that begins with all the things you are and ends with "I am a scientist" or "I am STEM." A student I worked with who was told by her family that she was an artist her whole life came up with the following: "I am curious, creative, thoughtful, and artistic. Because of that, I am a scientist." Now, she's a thriving Pre-med student at a well-respected institution after absolutely hating science in high school. She engineered herself into a STEM person. Writing your mantra down, printing it on a poster for your room, and saying it to yourself in moments of doubt will begin reclaiming who you are.

4. Find STEM people and make them your people

Most students who haven't found success in STEM don't know many (or sometimes any) people who are thriving in these disciplines. The only relationships to these subjects they have been through teachers who've either made them feel like they can't do well or who've been traumatized by their own negative STEM experiences. One powerful way to reconnect with STEM subjects—while also making necessary networking connections that will serve you well in college and beyond—is to reach out to those who work in these jobs. Many STEM professionals I know are interested in mentoring young students or describing their research to young people. Unfortunately, they don't know how to reach out to students, and students don't know they can reach out to them.

Do the work to find people who do the work

Think of an industry you love and try to find a STEM person who works there. Once you're aware of and forge relationships with the people in these professions, your outlook on the study area may change completely. Students I've worked with have found and built relationships with engineers at sneaker companies and math experts working on video games. They've earned internships and even paid jobs that began with research and sending an email to share their admiration and start a conversation. Building relationships with actual professionals demystifies these disciplines and reveals how STEM exists in much more powerful and exciting ways than you may realize. You'll learn from these experts about what they do and what it takes to work in a STEM profession. These amazing people in your network may also write you a college recommendation letter or offer sound advice about careers based on experience, not conjecture.

5. Experiment and embrace failure

When aspiring college students tell me about their awful experiences with STEM, I often say "congratulations" and then ask what kind of STEM work they've done. What have you built? What experiments have you done? What inventions have you dreamed up? What designs have you made? What broken objects have you fixed? They often look at me perplexed because they've never associated failure with something to be congratulated for. But in STEM, failing consistently is part of the process. Failing a class usually indicates that you did poorly on tests, but it does not indicate your potential for a future in STEM. If classes reflected the STEM discipline, the opportunity to keep retaking and redesigning the tests would be part of them. In the field, we embrace failure and consistently "fail forward." What we learn from each failure or misstep helps us improve and reach a more concrete conclusion. Failing is fun because it leads to figuring out what to do differently next time. If you failed a test, you failed at something someone else designed for you to judge your knowledge or intelligence—and it ends there. A true STEM student doesn't necessarily work well within that structure.

Make, design, create, fix, solve, and—most importantly—fail in order to sharpen your science-mindedness and STEM preparedness. Research experiments to replicate or find an old laptop to break apart and put back together. Then write about this experience in your college essays. This natural STEM identity is much more favorable to an institution than a student who only memorizes information but can't do anything with it.

Graduates with these attitudes, beliefs, and identities are in high demand, and STEM jobs have some of the lowest unemployment rates. These careers don't always involve working in a laboratory or having a fancy degree. They require college degrees in STEM or a related field and a willingness to think creatively and apply science-mindedness to what many call the "jobs of the future." Past failures or bad experiences shouldn't lock you out of reclaiming your STEM identity. These steps will get you on the right path toward amazing opportunities, rewarding careers, lifelong learning, and a brighter future.

6 DEVELOP A PLAN

Selecting the right career is not as simple as it looks. Mainly in high school, most students focus on finding the best college without even thinking about where they will land after completing college. It's never too early to start career planning, and having an idea about your future career allows a high school student to pick the rich college and course.

Every child starts to think about their career from the early seven, but they start discovering more about their career in high school. Before you decide about your college, a high school student needs to develop career plans and make it clear where they want to land.

Having a career plan allows you to be more focused, helps you learn about yourself, and offers ideas about how the industry works and the skills needed to enter the particular industry.

High school is the best time to get started with your career planning as it makes you aware of your weaknesses, strengths, skills, and knowledge required to achieve your future goal.

Once high school students enter college, they start to spend their time and money on achieving their career goals. That's why taking the right step toward career planning in the early days is important to determine the right career for your future.

Very few people are born with clear minds who know what they want to do and where they see themselves in the future. But most high school students are not aware of what they want to achieve in life; that's why it's important to plan things out so that you have true meaning and purpose.

Today many high school students rely on their interests when selecting a career or college. Knowing about your interests, skills, and academic achievement can help you during the career planning phase. Well, there are other steps involved that can help you concrete a more effective plan.

Below we will highlight the importance of career planning and how a high school student can get started with career planning to identify the best one for their future.

Why Is Career Planning Essential For Stem Students In High School?

Given that many urban students exclude Science, Technology, Engineering, and Mathematics careers from their career choices, the present study focuses on urban high school students and adopts the social-cultural approach to understand the following questions: how do students envision their careers? What are the experiences that shape students' self-reflections? And how do students' self-reflections influence how they envision their future careers?

Most of the students, after landing in their eighth grade, start thinking about what their future may hold. Many start exploring different career options they wish to land, and some students need guidance as they are unsure of the path. Career planning enables students to identify their strengths, skills, and interests that help them to determine the best career path for their future.

In every grade, high schoolers are asked what they wish to do when they grow up. It is one of the common questions every student comes across in the early days. Most students answer based on what they have learned (about a career) from their teachers, parents, or elders. On the other hand, some students make up answers or are unaware of what they want to achieve.

There are different surveys around the stats 52% of the students find the right career option only after joining graduation or at the end of the graduation. But in reality, high school is the best time to make the right career plan so that, depending on the plan, one can join the right course, get the right training, and build the right skills that can help them to land in the desired industry

In high school, when students start learning or exploring different career options, they change their minds multiple times after being introduced to a new career option. Not only that, but an individual can also change their mind after and enter the workforce. When new beginners learn about multiple careers, they are likely to switch as they find different career options that relate to their interests.

Limiting yourself to one career is not a good choice because high school is when you can explore different occupations, learn about different sectors, and many more.

High school is a critical time to identify your drivers for a properous career. Investigating different occupations and work settings will certainly fine tune your vision for post secondary school goals.

Career Planning for STEM Students in High School (Step-by-Step Process)

Strategic plans are tools that many organizations use to keep themselves successful and on track. A strategic plan is a roadmap for success. You can use the same plan to establish a route to academic success in high school or college. The plan may involve a strategy for achieving success in a single year of high school or for your entire educational experience. Ready to get started? Most basic strategic plans contain these five elements:

- Mission Statement
- Goals
- Strategy or Methods
- Objectives
- Evaluation and Review

Create a Mission Statement

You will kick off your roadmap for success by determining your overall mission for the year (or four years) of education. Your dreams will be put into words in a written statement called a mission statement. You need to decide ahead of time what you'd like to accomplish, then write a paragraph to define this goal.

This statement can be a little vague, but that's only because you must think big at the beginning stage. (You'll see that you should go into detail a little later.) The statement should spell out an overall target to enable you to reach your highest potential.

Your statement should be personalized: it should fit your personality and your dreams for the future. As you craft a mission statement, consider how you are special and different, and think about how you can tap into your special talents and strengths to achieve your target. You might even come up with a motto.

Sample Mission Statement

Stephanie Baker is a young woman determined to graduate in the top two percent of her class. Her mission is to use the gregarious, open side of her personality to build positive relationships and to tap into her studious side to keep her grades high. She will manage her time and relationships to establish a professional reputation by building on her social and study skills. Stephanie's motto is: Enrich your life and reach for the stars.

Select the Goals

Goals are general statements that identify benchmarks you'll need to accomplish to meet your mission. You will likely need to address some possible stumbling blocks on your journey. As in business, you need to recognize any weaknesses and create a defensive strategy in addition to your offensive strategy.

Offensive Goals:

I will set aside specific times to do homework. I will build relationships with teachers who write great recommendations!

Defensive Goal:

I will identify and eliminate time-wasting activities by half.

I will manage relationships that involve drama and threaten to drain my energy.

Plan Strategies For Reaching Every Goal

Please take a good look at the goals you've developed and come up with specifics for reaching them. If one of your goals is dedicating two hours a night to homework, a strategy for reaching that goal is to decide what else might interfere with that and plan around it.

Be real when you examine your routine and your plans. For instance, if you are addicted to American Idol or So You Think You Can Dance, make plans to record your show(s) and keep others from spoiling the outcomes for you.

See how this reflects reality? If you think something so frivolous as planning around a favorite show doesn't belong in a strategic plan, think again! In real life, some of the most popular reality shows consume four to ten hours of our time every week (watching and discussing). This is just the sort of hidden roadblock that can bring you down!

Create Objectives

Objectives are clear and measurable statements, as opposed to goals, which are essential but indistinct. They are specific acts, tools, numbers, and things that provide concrete evidence of success. If you do these, you'll know you're on track. If you don't accomplish your objectives, you can bet you're not reaching your goals. You can kid yourself about many things in your strategic plan but not objectives. That's why they're important.

Sample Objectives

- Buy a planner and write in it every day.
- Sign a homework contract.
- Secure a device for recording my favorite shows.
- Take a learning style exam to determine my best learning style.

Evaluate Your Progress

Writing a good strategic plan on your first try is difficult. This is a skill that some organizations find difficult. Every strategic plan should have a system for an occasional reality check. If you find, halfway through the year, that you are not meeting goals, or if you discover a few weeks into your "mission" that your objectives aren't helping you to get where you need to be, it may be time to revisit your strategic plan and hone it.

When Do Students Begin Thinking About Their Career Interests?

Most of the time, career planning begins after a student lands in the final year of their under graduation or post-graduation (when they are searching for a new job). Well, college students can still identify various options during that point in time, but it's considered too late to find the right career for their future. During the last hour, they select a career option they don't wish to enter. Decide when to start thinking about their career planning? Write it down.

According to recent research, students gain career-related and occupational interests as early as the seventh grade. That is when they start determining the importance of career goals and what they wish to be. As students start upgrading to the next grade, they become more curious about their future goals, which is when they should start planning their careers.

In the early days, students start to select career choices depending on their interests and values, but as the day passes, their career planning becomes more dependent on how much money they will make or whether they have a good job at a reputable company or not. Instead of focusing on the job and paycheck, I recommend students select a career that aligns with their interests, values, and, most importantly, skills.

During high school interests of every student regarding their career development are at the same rate; these interests generally include thinking about having a technical or scientific career like becoming an engineer, teacher, doctor, and more.

Most researchers agree that these interests start to develop as a result of the environment they are living in. High School is the best time to help students get exposure to different industries and participate in different activities to select the right career path for their future. Students who have explored various career options are more informed about their career planning.

Is There Any Relation Between Student Academic Achievement And Career Interest?

Undoubtedly a student's career interest and academic achievement overlap; sometimes, academic achievement helps students to determine the right career path, whereas there are times when academic achievement doesn't match with career achievement.

For example, a student may have gained good achievement in a particular skill or subject but don't feel excited when selecting it as a career goal. On the other hand, a student may be interested in a particular field, industry, or subject but have few skills or knowledge.

While making a career plan, students should focus on both skills and interests. Remember, skills can be developed over time, but your interest should relate to the career option. Most of the time, people do a job in an industry that doesn't excite them and makes them feel satisfied.

Most high school students may not find themselves good at the skills required to enter a particular industry. But it can be developed with the right strategy and by creating a short and long-term plan. Students will find their interests stronger than skills, but skills can be developed once you decide which industry you want to enter. To select the right career plan, go behind your interests and check whether you can accomplish those skills. Depending on that, you can find the right career option.

Are High School Students' Career Interests Consistent With College Majors?

Today, students must have interests consistent with their college major. Most students enter higher education with strong ideas, structured career options, and a sense of confidence to accomplish their goals. But they start losing interest after landing the desired college major, which directly affects career planning.

Researchers say that students who are familiar with their interests and keep their interests consistent with their choice of college majors are likely to have more structured career goals. The same goes for the job; students interested in their job will have a more satisfying experience.

There are times when student interests are consistent, but somehow the career option they select may not be consistent due to current and future workforce demands. For example, most students prefer to join occupations like engineering, visual art, applied art, and more, but sadly due to the huge crowd working for the same goal, the demands are comparatively less. Whereas your desire to enter into industries like agriculture, nursing, pharmacy, and more. Those are areas where demands are high, but job seekers are fewer.

If you don't want your interest to die as you move ahead in your career, then we would recommend you make an education and career decision that allows you to keep your interest consistent.

High School is considered one of the most crucial times to start career planning. Above, we have listed all the tips on how high school student can start their career by exploring different industries. High school is the time when students get more curious about knowing different industries. Using our highly advanced AI-driven career assessment platform, students can not only know about different industries but also explore each one individually.

8 STAY FOCUSED

Focusing on high school classes can be difficult if you don't love every class, and you might find that it's more interesting to talk with friends or keep up with your favorite television shows and social media accounts. Focusing at all right now, during a global pandemic, can also feel impossible.

However, it's important to stay focused in high school so that you can learn what you need and graduate with the credentials you want to help you reach your future goals.

Tips On How To Stay Focused In High School

Whether you are learning in person this year, online, or some hybrid of both, here are 10 tips on how to stay focused in high school. And actually, these tips will also help you in college and beyond.

1. Identify your learning style.

One of the first steps toward knowing how you can stay focused in high school is to identify your unique learning style: auditory, visual, tactile/kinesthetic..

Are you a more auditory learner and prefer to hear a lesson out loud? Or are you a more visual learner who needs to see words on a page or screen to understand a concept? Alternatively, do you learn better when you can apply lessons to hands-on experiences?
To identify your learning style, you can take an online quiz, ask your parents and teachers, or just sit with yourself and think about how you usually like to learn.

If you notice that you have responded better to certain lessons in school, take note of how those lessons were taught. Making an inventory of successful learning moments will help you recognize your style of learning and also find ways to help augment your learning as a result.

Once you understand how you learn best, you can try to position yourself to learn this way. Even if your teacher doesn't teach the way you like to learn, you can create ways to review material that activates this type of learning.
 For example, if you have to watch a lecture and hear a teacher's words, make sure you also read the textbook or write notes as a visual learner to see the lesson's content as well.

You might also let your teachers know what you believe your learning style to be. They might not be able to accommodate you individually every time, but they will likely try to help however they can.

2. Actively participate in class discussions.

Another great way to stay focused in your high school classes is to actively participate! This might seem obvious, but when you take an active role in class discussions, you will take more ownership over the material you're learning. You will also be more likely to remember information and want to pay attention in class.

Some high school students will be in physical classrooms this year, while others will be online. Either way, make sure you take as many opportunities as possible to actively participate.

This will likely help your grades, since participation is often part of those grades, but it will also help you stay engaged. As a result, you will enjoy your experience much more, and classes will speed by.

Even if you're normally shy, try to challenge yourself to participate however you can. Try writing down ideas that you think of while your teachers or classmates are speaking, so you can read your notes if you're nervous about speaking in public. Writing notes will also help you remember material and make your own personal connections to it.

3. Create a productive learning environment for yourself.

Whether you are at home, at a neighbor's house, in a community center, or in a classroom environment to take high school classes this year, make sure to create a learning environment for yourself that will be productive.

Of course, you don't always have a lot of control over your learning environment, but exert whatever control you can.

This might mean you sit near the front of the room in the classroom, make sure you aren't sitting next to a friend who will distract you, or situate yourself in front of a window to make you feel more comfortable.

You can also keep inspiring quotes or photos around you to remind you to pay attention as you learn.

Make sure that, even if you're at home, you have a chair that allows you to sit upright and is comfortable but not so comfortable that you want to nap. Do whatever you can to keep your learning area clean and organized.

All of these tactics will help you to be more productive and pay attention in high school, either in an in-person or online format.

4. Take notes.

As previously mentioned, taking notes will help you to remember material and also allow you to connect personally to the material.

However, don't just take notes to take notes. Find a way to take notes that makes sense to you so that you can read them later and understand what you've written.

Make diagrams, write in bullet points, or use color-coded pens. You can even draw pictures if that helps you understand and remember content.

However you take notes, you should find a way that works best for you and will allow you to use your notes to review material for tests at a later date.

5. Preview and review material.

Previewing material before class and reviewing material after class will also help your focus in high school. If you know what your lesson is going to be about, consider looking over any classroom material prior to your class or even doing some of your own research online. This will prime your brain to learn and understand the material and might even give you some talking points for class discussions.

Another strategy is to review material you've learned right after class is over. This tactic will help you solidify that information in your mind and allow you to process what you've learned. This doesn't have to take long, just five minutes or so after a class, but you'll be surprised how effective it can be.

6. Keep a calendar for assignments and deadlines.

While you're in high school, it's easy to rely on parents or teachers to help you remember important deadlines. But you will focus much better if you keep track of all of that information on your own.

Keep a calendar, whether on an electronic device or in a physical planner, to help you remember when assignments are due. This will train you for your future as a college student and/or employee and will help you take control of your own destiny while in high school.

If you use your phone to keep track of deadlines, you can even set reminders to study or review material. These reminders will help keep you on the right track while you're wading through your high school lessons.

7. Eliminate distractions.

In a digital society, eliminating distractions seems impossible. However, if you want tips on how to focus in high school, this is probably the most important one to consider as a starting point.

Though you might think you're multitasking when you text a friend or check social media while learning, studies have shown that you can't actually multitask.

Instead, you can only focus on one task at a time. So, if you receive a text or social media notification during class, your attention will inevitably be diverted to that notification.

To avoid these types of distractions, keep your phone out of your learning space. Turn off social media notifications during class. You can even download anti-distraction apps to block distractions from ever getting to you in the first place.

8. Practice healthy habits.

Keeping up with your health is an important area of focus no matter how old you are or in what year of school. When your body is functioning properly, your mind will focus better, and you will be better equipped to stay on task during your high school classes.

Get plenty of sleep, drink water, and eat healthy meals. Take time in hours when you're not in classes or other activities to exercise. Meditate if you need to quiet your mind.

Essentially, follow all the healthy habits that you've heard about over time. Though making healthy choices does not always feel fun, your body and brain will thank you. And, you will definitely be more focused in high school.

9. Make connections.

You will also stay more focused in high school if you make connections.

Make connections between your lessons and the world around you. Find ways to relate class concepts to your own life or examples in the news. This will help you remember material and also likely spark your passion for a topic. Learning should not just be about getting high grades and preparing for a future in college or the workplace; learning should also be fun.

Make connections with your teachers and classmates as well. Especially during a global pandemic, personal connections are important. Remember that your teachers and classmates are human beings with feelings, fears, and goals as well. This will help you enjoy the high school experience much more and allow you to lead with empathy as you participate in your classes.

10. Keep your future goals in mind.

Finally, you will inevitably focus much more in high school if you keep your future goals at the forefront of your mind. Why does doing well in high school matter to you? What do you hope to achieve when you graduate? Do you have a target college or internship?

Write down your future goals and keep them close to your work space during classes. These goals will remind you why you want to focus on high school, even when that can feel difficult.

Staying Motivated In STEM

Spending hours on the microscope counting cells, having a failed experiment, or getting brutally honest feedback can bruise your confidence and blunt your excitement for science. Making sure you've built up a reserve of positive feelings about your research will help you power through some of the low points. Below are a few tricks that I've found helpful to spur me on when I was frustrated, tired, or even scared. While these tips might help manage minor dips in motivation, if you are experiencing crippling self doubt or a total absence of motivation, please talk with someone you trust!

Keep a Victory List

As grad students striving to be our best, we tend to be very hard on ourselves and focus heavily on our perceived errors. Take the time to consciously face down your inner critic. Record anything you accomplish that gives you that "good job" rush. Write down challenges you've overcome, awards you've won, times when you were able to help someone you respect, positive feedback you receive, or anything that made you feel smart and productive. Be generous with yourself. These can be silly victories, like getting an $R^2=0.999$ on a Bradford assay standard curve (nice pipette work, my friend!); personal bests, like giving a talk without panicking; or very tangible achievements, like having a paper accepted. You don't have to show anyone this list. Just make sure to keep it handy so that you can update it often with the great things you've done.

When the bad things happen or when you're facing a big challenge, it's easy to forget all about your awesome accomplishments. You might start experiencing impostor syndrome, thinking that maybe you're a fraud and don't belong in grad school. Looking back at your victory list can help remind you that you are smart and successful, and you are supposed to be here. Because you are.

Re-visit Your Goals

While the victory list looks back at your past successes, revisiting your goals looks forward to your next achievements. Ideally, you will have a few big dream goals. Each big goal should then be broken down into medium-sized ambitious goals. Finally these medium-sized goals should be broken down into achievable stepping-stone goals. When you see that those frustrating experiments you have to run are part of achieving your big goals, it's a lot easier to convince yourself to stop looking at Facebook, walk to the lab, and get going.

Make sure you have a hierarchy of goals for both your personal and professional life. It may be hard to find a motivating starting point in your professional life, so if you can attack a goal in your personal life, go after that for a time. Once you get some momentum in working on your personal goals, you'll be better able to keep rolling forward toward your professional ones.

Re-visiting your goals also might help you realize why your motivation is dipping. It could be that what you are struggling to get working on isn't aligned with your big dream goals anymore. It may be time to re-assess things more seriously, with the help of your mentors and trusted friends.

Listen to Inspiring People

There are many great (free!) podcasts featuring scientists talking about their work. Oftentimes, they talk about not only their successes, but also the challenges they've overcome. The passion these scientists have is highly infectious. Hearing that they've faced down similar challenges is a good reminder that discouragement is normal and can be overcome in many ways. You can listen on your computer or load podcasts onto your mobile device.

Four of my favorite podcasts are:

The People Behind The Science. Dr. Marie McNeely asks established professors about their research, hobbies, favorite books, biggest failures, motivating forces, and advice for young scientists.

The Life Scientific. In this BBC podcast, Jim al-Khalili interviews eminent British scientists and learns the stories behind their greatest findings.

Breaking Bio. This podcast is hosted by a crew of graduate students and post-docs. It often has interviews or discussions with other graduate students, researchers, or science communicators. It's much less formal than the other podcasts mentioned here, but has some of the funniest and most relatable discussions.

Quirks and Quarks. Bob McDonald hosts this science news radio show on CBC. Three or four researchers are interviewed in each episode about their new papers. This podcast is focused on understanding the implications of their findings, but also delves into how difficulties were overcome to obtain important data.

Do Outreach

Getting involved in outreach can help re-kindle your love for your research. The work you do every day, which feels mundane and boring, will seem very exciting when you explain it to the public. There are lots of on-campus opportunities to volunteer in formal science outreach organizations (e.g., Let's Talk Science in Canada). Off campus, volunteering at a museum, zoo, or conservation area related to your research might also be an option. Don't shy away from working with little kids, either. They are often the most enthusiastic about science. You can do outreach in a less structured manner by explaining your work to your non-sciencey family and friends. Hearing yourself talk about your research and how it fits into the bigger picture will also remind you that your work is important.

Actively show yourself that you are working toward your dreams, and that you are successful. Realize that everyone has ups and downs with research, but remember that your research, and science in general, are exciting and inspiring. In doing this, hopefully you will be able to push through some of the low points with a little less struggle.

All the work you're doing now will be worth the effort in the future when you reach your goals! That should be motivation enough to stay focused. You've got this!

8 THE MIND, BODY, SOUL APPROACH

There is strong believe that academic and personal growth remain incomplete unless the mind, body, and soul are fuelled. Keeping this approach in mind, in this chapter, we have developed our curriculum and the routine of our students in such a way that all these 3 facets of Human development are ignited and nurtured on a day-to-day basis.

MIND

We believe in empowering our STEM students by helping them learn to learn. And this empowerment will not be limited to their school life alone but will stay with them throughout. We strongly want the natural intelligence of our STEM students to be sharpened and preserved.

Less Stress: For educators, mindfulness can decrease burnout and foster wellbeing. When taught to students, mindfulness supports more relaxed and regulated learners.

Emotional Regulation: Mindfulness supports happiness and emotional balance for both educators and students while developing resilience to the numerous pressures of live.

Building Focus Muscles: Mindfulness strengthens attention and limits distractibility, helping us stay attuned to what is happening in the present moment and building productivity.

Culture of Compassion: Mindfulness supports the development of kindness and understanding while limiting bias and assumptions.

Gain Expertise: Train with experts in the mindfulness and social and emotional learning fields and learn best practices for sharing these

Connect to Community: Join and inspiring community of fellow educators committed to compassionate teaching.

BODY

Learning through sports is not just part of our education methodology, it's another main stream of social and practical education.

We use sports as a tool to build skills, developmental assets and a wide range of values including fairness, team building, equality, discipline, inclusion, perseverance and respect.

Sport has the power to provide a universal framework for learning values, thus contributing to the development of soft skills needed for responsible citizenship. We believe in learning through sports and skill building through sports

SOUL

We want children to be healthy and happy, not just now but for the rest of their lives. And teaching them the art of meditation at such a tender age will help them stay calm, happy and in a healthy space of mind.

Meditation helps strengthen moral purpose and ability to make the right (and sometimes tough) decisions.

Creating positive beliefs about what can be achieved helps students realize their own strengths and capabilities.

To nurture the soul, we also try embedding new positive behaviors that support the development of excellence in oneself and others.

Through a range of leadership workshops focusing on resilience, positive-psychology, inter and intra-personal relationships and our values, we help students let go of the false belief that well-being is optional and only something to be considered when there is crisis.

Students are encouraged to believe that self-care and attention to their inner most needs are integral to being able to thrive as school leaders.

By nurturing these three facets of human existence, we try developing individuals to become well-rounded leaders of tomorrow.

Simple Ways to Balance Your Mind, Body, and Soul

True balance is more than just a balanced diet or a balanced work life. It's not just about how many hours you sleep or how long you meditate every day. True balance is a state of mind–a balance of the body, mind and heart. It's about finding the perfect harmony in all aspects of your life. It's about finding the sweet spot between the unbalanced things about ourselves that make us feel sad, angry, lost, sick, or unfulfilled. It's about finding a way to feel happy with the way you are, without obsessing over the things that make you feel that way.

When we think of health, nutrition and exercise are usually the first things that come to mind. However, good health is not just about the physical body. Our mind and body are connected and have a great influence on each other.

There are many things you can do in your daily life to achieve overall wellness. Here are 25 simple ways to begin cultivating a balance between mind, body and spirit.

1. Read and study often. Your education may not end after you leave school. Open your mind to new possibilities, beliefs and interests by reading, taking online courses, watching documentaries and attending workshops.

2. Meditate regularly. Meditation improves memory, attention, mood, immune system function, sleep and creativity. It only takes a few minutes a day to reap the benefits, and you can start with a free 30-day meditation course. Guided meditations are ideal for beginners.

3. Practice yoga. Yoga is surprisingly good for your overall health. It helps develop your strength, coordination and flexibility while calming your mind. It also includes the relationship between mind, body and spirit.

4. Avoid long periods of sitting. Try to stand or move around while you work, if possible. Sitting too long is associated with heart disease, diabetes and lower life expectancy.

5. Take at least 15 minutes of moderate to vigorous exercise every day. Do you live near your workplace? On a good day, hike or bike. Physical activity is important for heart health, stamina and mood.

6. Spend time outside. This is the perfect time of year for camping, boating, picnicking, outdoor sports, wild food hunting, camping and more!

7. Incorporate more plant-based foods into your diet. Eating plenty of fruits and vegetables can help prevent chronic diseases. Buy fresh, seasonal produce at your local farmers market.

8. Get involved with a volunteer organization or activist group. Use your voice or your talents for the good of the world. We're all connected, and it's incredibly satisfying to feel that connection when working toward a common goal.

9. Reawaken your passions. Take time each day to do what makes your soul happy. Many of us work so hard that we forget how beautiful it is to draw, dance, compose music, write, garden or swim.

10. Listen to music often. And sing or dance along!

11. Be grateful. For example, take time each day to write down or

think about what you are grateful for. B. for family, friends, pets, food, shelter, health or the beauty of nature.

12. Be friendly with everyone. And that goes for you too!

13. Get a good night's sleep every night. And don't forget it: You're never too old to sleep.

14. Detoxify your beauty care. Switch to natural products.

15. Remove harsh chemical cleaners from your home. Buy eco-friendly cleaning products or make them yourself.

16. Find a career path that makes sense to you. Pursue your dreams, not your wealth.

17. Let go of the little things. If something doesn't matter tomorrow, don't let it ruin your day.

18. Take it easy. A little rest and relaxation, if you are used to being on the road a lot, can restore your mind and body.

19. Stop trying to please people. There's a difference between being friendly and being a doormat. If you worry too much about what others will think, you will lose yourself and end up feeling unhappy.

20. Eliminate the main sources of stress in your life. Like wasteful spending, clutter, a lousy job or unhealthy relationships.

21. Avoid gossip. Condemning your neighbors and co-workers doesn't make you superior, but it does make you hard to believe.

22. Laugh a lot. If you take life too seriously, you'll miss a lot of beautiful moments.

23. Travel and discover other cultures. Do it as often as possible!

24. Forgive yourself for your past mistakes. Learn from the past, but don't let it destroy you.

25. Use natural remedies whenever possible. Under the guidance of an alternative practitioner, herbs, good nutrition and essential oils can be very healing and have less dangerous side effects than most pharmaceutical drugs.

Which of these suggestions would you like to implement in your daily life?

You've probably heard the phrase "mind, body, and soul." The mind is the source of all our thoughts, feelings, and actions. The body is the source of all our movements, as well as our organs and senses. The soul is the source of all our experiences, which means our thoughts and feelings as well as who we are..

9 YOU HAVE THE POWER

"The power is yours!" Its one of my favorite quote from the popular cartoon Captain Planet. As an engineer, I am often task with solving problems and then moving on to a new problem. My initial belief before tackling any issue to be believe I am capable of finding a concrete solution. It all starts within!

Many of the recommended tools and ideas shared throughout this book will guide readers in the direction of technical preparedness. What I truly want each reader to FEEL is empowered. Empowered to take on difficult information, focus on your reason for pursuing your endeavors and see it till the end.

The definition of power according to the laws of physics states, "the amount of energy transferred or converted per unit time." This is important to understand because the it's the drive and motivation and dedicated effort over time that empowers potential engineers, scientist and doctors to become notable professionals.

The world we live in is changing, and we must keep pace with it. STEM education changes society by offering learners a new mindset and skills valued in any profession. They allow young people to be flexible, look for patterns, find connections, and evaluate information. Besides, STEM education raises social awareness. It communicates global issues to the general public. Therefore, STEM opportunities move us to a knowledge-based economy and enhanced sustainability literacy.

STEM education goes beyond school subjects. It gives a skill set that governs the way we think and behave. Merging science, technology, engineering, and mathematics, STEM education helps us to solve the challenges the world faces today. Using the methods, resources and techniques discussed in the text will ensure your steps are Confidently Calculated!

www.ingramcontent.com/pod-product-compliance
Lightning Source LLC
Chambersburg PA
CBHW071417210526
45465CB00001B/432